Management Division 2023

AA002886

Held at the 2023 AIChE Spring Meeting and 19th Global Congress on Process Safety

Houston, Texas, USA
12-16 March 2023

ISBN: 978-1-7138-7924-4

Management Division 2023

Held at the 2023 AIChE Spring Meeting and 19th Global Congress on Process Safety

Houston, Texas, USA
12-16 March 2023

AIChE
The Global Home of Chemical Engineers

Printed from e-media with permission by:

Curran Associates, Inc.
57 Morehouse Lane
Red Hook, NY 12571

Some format issues inherent in the e-media version may also appear in this print version.

Printed with permission by Curran Associates, Inc. (2023)

For permission requests, please contact AIChE
at the address below.

AIChE
120 Wall Street, FL 23
New York, NY 10005-4020

Phone: (800) 242-4363
Fax: (203) 775-5177

www.aiche.org

Additional copies of this publication are available from:

Curran Associates, Inc.
57 Morehouse Lane
Red Hook, NY 12571 USA
Phone: 845-758-0400
Fax: 845-758-2633
Email: curran@proceedings.com
Web: www.proceedings.com

TABLE OF CONTENTS

ROLE OF LEADERSHIP IN PROCESS SAFETY IN EVOLVING ORGANIZATIONAL CULTURE

MANAGING TO DELIVER CORPORATE SUSTAINABILITY COMMITMENTS

18c *From Fear to Understanding: moving Process Safety Focus Towards Value Creation and Profitability*

Jimmy Miller

The OSHA Process Safety Management Regulation was introduced in 1992. It was developed to help plants, process, and people to be able to work more safely. However, the resulting message has often been misunderstood, with greater focus being applied to avoiding fines than generating better plant safety, thus creating a culture of fear, rather than improvement. Regulatory compliance is of course critical, but its application without consideration may not read through to true safety and operational performance improvement, or to the benefits of greater profitability that can result from a well implemented system.

This presentation explores the history of process industry incidents and their causes while drawing parallels to the early days of the automotive industry and their own struggles with safety. Much can be learned from that evolution, and it mirrors many of the same challenges we face regarding process safety. By utilizing inherently safe design concepts with better capturing of operational data including increased diagnostics, safety and business decisions can be made more quickly and efficiently. This leads to inherent regulatory compliance and increased business improvement.

18b Inclusion in Process Safety: What Should You Know?

Jordan Bailey, Chason Coelho

How do the objectives of inclusion factor into the theory and the practice of process safety? What can be done to achieve process safety objectives while also developing an inclusive workforce? What are the tangible benefits to incorporating inclusion elements into process safety programs? This talk addresses these and other impactful questions. It focuses on the intersection of inclusion and process safety at two levels. One level provides important information on issues such as the prevalence and types of disabilities in populations that may work in the process industries. The other level provides actionable insights into how (a) accommodations can be made for individuals who have disabilities without sacrificing process safety performance and (b) the knowledge, skills, and abilities of such individuals can promote process safety performance.

18a Closed-Loop Model to Optimize Management of Change System and Best Practices to Manage the MOC Digital Transition

Hussain Alabduljabbar

Management of Change (MOC) is one of the key elements in any modern model of a safety management system. Due to the dynamic nature of the oil and gas industry, frequent changes are essential to maintain facilities and optimize production. Changes come with associated risks that require effective management to ensure safe and effective execution. The MOC system includes identifying, reviewing and authorizing changes that are not replacement-in-kind to ensure any potential adverse impact is addressed. The system also involves documenting and communicating changes to affected employees. Failure of the MOC system has been considered as a major contributor to incidents in the industry. Both operational and organizational changes, if not managed well, can lead to major industrial incidents. For example, improper control of changes has been identified as a major contributor to the Flixborough disaster in 1974, and Texas City Refinery explosion in 2005. Improving the implementation of MOC requires a healthy system in place and adequate levels of competency. In addition, digitalizing the MOC system can enhance effectiveness of risk management when the integration and transition, from manual to digital practices, are managed effectively. The paper proposes a closed-loop model to systematically overcome challenges related to the implementation of the MOC process within the oil and gas industry, with provisions to ensure prioritised continuous improvement. Additionally, it discusses the digitalization challenges and focus areas to ensure effective deployment.

The failure of the MOC system can have different causes. The main cause is the failure to trigger the MOC process to manage applicable changes. This is still being identified as a contributory factory to major incidents in the industry. Other challenges include failure to implement adequate hazard identification and risk assessments, failure to revalidate control measures as part of temporary changes, and failure to update process safety information. A closed-loop model (between the technical authority and the auditing function) will systematically drive prioritised improvements with fit-for-purpose safety tools.

In addition, emerging technologies and digital solutions for MOC can enable better management of risks by minimizing human errors, improving efficiency, and simplifying the process implementation. Likewise, big data and advanced analytics can uncover invisible patterns and correlations. Like any other digitalization change, digitalizing the MOC process comes with its own challenges that could lead to unintentional results such as overwhelming the operator with unnecessary alarms/notifications, losing track of objectives due to the visibility of large volumes of data, missing the opportunity to drive effectiveness with unconventional Key Performance Indicators (KPIs), and losing the required technical expertise as a result of automation.

88a Delivering Corporate Sustainability Commitments Managing Collaboration and Urgency

Gayle Schuller

The kick-off talk will focus on best management approaches to foster and achieve a corporate culture which centers in sustainability, meeting and delighting stakeholders (customers, employees, shareholders, community members, etc.) expectations by managing collaboration with a sense of urgency. We will discuss how have tremendous opportunities to create new and improved materials with desired sustainable chemical and physical attributes.

88b Sustainability Considerations in Process Technology Developments

John Dever

Corporate Values and Tenets of Operation impact the culture of an organization as it meets its objectives for the Triple Bottom Line (TBL): People, Planet, Profits. For the chemical industry, process technology and design have a significant role in meeting the Sustainability goals that are woven through the TBL. Waste, emissions and process safety are key concerns to employees and the communities that reside in close proximity to chemical production facilities. Process designs will dictate the energy efficiency, process yields and selectivity that determine emissions and govern the financial bottom line. Corporate Leadership can foster a Sustainability culture by expecting, encouraging, and supporting process design considerations that have positive impacts on Sustainability. Examples of Sustainability considerations in design include continuous processing to reduce waste and inherent safety practices to improve process safety. This presentation will discuss how AVN Corporation incorporates Sustainability considerations in the development of process technology for all customers, with an example, as well as a discussion of how AVN's own operations are impacted by a TBL mindset.

88c Driving Sustainability Progress: The herding of cats?

Darrell Boverhof

The Sustainability/ESG landscape is rapidly evolving with demands and expectations coming from many different stakeholders. This environment can be overwhelming and create complexity in achieving a clear company understanding of sustainability priorities and actions for advancing meaningful progress. This session will discuss the current complex landscape and provide perspective on how to create internal alignment that drives clarity and progress while satisfying stakeholders.

88d From an Environmental Liability to a Sustainable Asset

Ashley Saunders

Just outside the small town of Oak Ridge, Tennessee the reuse of a large, government industrial site is contributing to positive environmental, economic, and societal change. A massive environmental cleanup project led by the Department of Energy and UCOR has transformed nearly 2,200 acres that once supported national defense and energy missions from an environmental liability to a community asset. The removal of 500+ contaminated buildings and thousands of cubic yards of impacted soils has improved air, land, and water quality. Throughout the cleanup journey, sustainable practices have been at the forefront of project execution, including reuse of electrical and other equipment; sustainable purchasing/materials programs; and waste diverted via recycling. The cleanup effort has also resulted in ecological restoration and contributed to a better quality of life for nearby residents. The once blighted property is now a growing industrial park with greenspace and natural areas that provide public recreational spaces, birding, and wildlife viewing opportunities. Solar fields dot the landscape and clean energy businesses are starting operations and bringing new jobs to the area. It is also home to the nation's 409th National Historical Park dedicated to education and preservation. This success story provides an example of how turning an environmental threat into a considerable asset is ensuring a sustainable future for the East Tennessee region.

AUTHOR INDEX

NOTES

NOTES

NOTES

NOTES

NOTES

NOTES

NOTES

NOTES

NOTES

NOTES

AIChE
120 Wall Street, FL 23
New York, NY 10005-4020

ISBN 978-1-7138-7924-4
9 781713 879244